Bluebell

ZINOIC RAYS, THE SECRET TO ALIEN INVASION ON EARTH

Tobechi Onwuka

Zinoic rays, the secret to Alien Invasion on Earth
1/1/2018

Dedication

'This piece of work is for all lovers of science fiction stories, tales of Alien invasion, horror, mysteries and suspense'. Have you ever wondered why sometimes you see strange finger prints on your glass window?, or your dog keeps barking at an unseen stranger?. What about humans that behave weirdly?, this book may have the answers....keep reading.

CONTENTS

CHAPTER ONE

ORWELL

Sarvile drive is a small estate in the town of Novraska where a few staff and members of the famous Ralph Research Institute, Novraska, reside. As the rains started dropping, Orwell smith rolled down the glass slides of his apartment. Orwell is a virologist in Ralph research Institute where he had been working for close to eleven years. The 35 years old Orwell began his career early in life and had worked with several institutions as a lab scientist before he was eventually employed by Ralph Instute. Orwell has few friends, among whom is James who also works at Ralph research institute as a lecturer. Both of became friends back then during their days in the college. Coincidentally both friends happened to meet again as employees at Ralph Research Institute. Ever since then, they have been birds of the same feather, though both of them are still single living in the small town of Novraska.

Novraska is a small town mainly dominated by migrants from Allaport, Swariviva and Demark. Though many believe that Novraska has come to be a no-mans-land for most of the dwellers. One unique feature of the town is that, though it seems to be calm and cool but has in the past was said to have recorded the highest cases of UFO appearances on Earth. For somebody like Orwell, who just moved into the town about a year ago believed that all those things said about Novraska were all a fairy tale to him.

On weekends, Orwell normally hangs out with few of his next door neighbors, like Mr Lake gari, who keeps lots of pets with him and his family. Sometimes Orwell would spend great junk of his time playing with Mr lake's moppy dog each time he visited him and his family. Moppy enjoys playing with Orwell to the extent that he would sometime hop down to Orwell's apartment to play with him. One nasty habit Moppy had was that he easily barked at anything that looked strange to him.

One of such cases includes the day Annie Thomson, Orwell's Swedish girlfriend, visited Orwell on a weekend. Annie came putting on dreadlocks as hair style. As soon as moppy met Annie at the door, he barked aggressively at her to the extent Annie, in an attempt to run away, crashed into the hedges in front of Orwell's apartment before Orwell intervened. Orwell immediately apologized to Annie but already the deed had already been done. Orwell took the dog back to Mr. Lake, his neighbor and complained to him some strange attitudes of the dog, more especially, his unexplainable aggression and constant barking at space. Mr. Lake gave him no tangible answer rather he expressed fear that the dog might be,' possessed or something, he', he replied back at Orwell.

CHAPTER TWO

STRANGE

On a Saturday's evening, Orwell and Moppy embarked on a walkout to James's apartment, located at the west wing of Sarville drive. Along the way, Orwell noticed that Moppy stopped at every signposts and trees along the walk way to James's apartment. Sometimes, he starred vaguely into space and barked at nothing. Orwell observed the dog's weird behavior as they walked and wondered what the dog could be starring and barking at. As they got to a point very close to James's apartment, Moppy suddenly stopped, starring frantically into space. He sometimes drew back a bit in readiness to combat with whatever invisible thing that provoked him. Orwell at that point became confused because he could not see what the dog was furious about. Sarville Estate is quite a calm Estate compared to other Estates found in Novraska town and so couldn't find any reason for the dog's sudden aggression.

Orwell eventually succeeded in calming down the dog while on the walk way, he speculated that whatever might had actually made Moppy this upset must be something really 'strange'. Orwell wished there could be a way humans could converse freely with animals in their ecosystem. When Orwell got to James's apartment, he narrated his strange experience to his friend. James affirmed to the fact that sometimes strange things do occur in the town. He also told Orwell of his recent experience with something he couldn't explain. James said he saw three finger prints on his glass window late in the night, four days ago. None of the three finger prints resembled that of humans nor any animal known to him. All efforts made to capture the strange finger prints proved abortive, as the prints could not appear on any camera lens. After James had narrated his own strange experience,

Zinoic Rays, the secret to Alien invasion on Earth

Orwell came to the conclusion that there could be strangers out there that no one knows who they are.

CHAPTER THREE

ORWELL MAKES A NEW DISCOVERY

Early the next morning at Ralph research Institute, Orwell walked into the science laboratory where he works. At the moment he happened to discover a new kind of virus that he got from the bark of an ancient tree in the provincial areas of Novraska. Orwell got thrilled by the unique communicative ability of the virus under the microscope. Orwell wondered how a microbe could have such unique communicative impulse. Orwell took extra time and care to further study the composition, characteristics and instrinstic features of the newly discovered virus.

Orwell got fascinated so much that he spent the next four hours thinking of how possible to interact with the virus. Orwell"s curiosity and inquisitiveness led him to wonder how man can communicate verbally or in the form of a decoded text with other members of his ecosystem, more especially the animals?. Orwell speculated that in the nearest future, especially with the technological advancement in global communication, man may be able to interact freely either verbally or through text with animals and even microbes. He speculated that most of these microbes had been existing long before humans and therefore most of them may have a clue or assist humans unlock the mystery behind their existence on Earth and the universe at large. For over three days, Orwell dedicated his time to intensive research on how possible to interact with living microbes. The more he searched and researched, the more he found clues to satisfy his curiosity, until he finally found and built the first man-ever-made microbe's intelligence decoding device.

Zinoic Rays, the secret to Alien invasion on Earth

The device uniquely decodes and transcribes communicative movements, impulses and intelligence of microorganisms. Orwell named it MID, meaning, microbe's intelligence decoder. The device has a text screen, a keyboard and internal hypersensitive sensors with some decoding circuits. Orwell geared up to launch his device with the newly discovered virus; he attached the tube where the virus was kept to the device and sat back to see what would happen afterwards.

CHAPTER FOUR

SOMEONE COMES ON BOARD

Ten minutes gone and nothing showed up. Orwell, though optimistic that something good may come out of his four days intensive research, left the laboratory for a quick lunch at the college's staff restaurant located southeast of the Institution. Orwell met James there in the restaurant. The two had their lunch together, but Orwell never told him about his latest project for some reasons. Orwell knew James too well, he could hardly keep a secret and sometimes turns out to be a wet blanket. Instead Orwell told him about Annie and how they have been getting along for some days. James felt excited to hear that his friend, Orwell, has been dating a lady for once in his life. James knew Orwell hardly had time for women. At a point, James thought it wise to recommend his 23 years old next door neighbor, Bella, to Orwell. But that certainly didn't work out because Orwell never for once fancied that idea.

Immediately after lunch, Orwell walked back to the laboratory. On getting there, what he saw on the device's text screen baffled him. For about three minutes, Orwell stood in awe, gazing at the text screen which had the word, 'Hi', on it. Orwell looked through the microscope and saw that the virus was even more excited than himself. Orwell quickly summoned enough courage and replied back by typing in the word, 'Hello'. Orwell at this point got fascinated, he waited anxiously for about ten minutes before the next text came in saying, 'who are you?'.

Out of excitement, Orwell replied, 'I'm Orwell and you?'. The virus also replied back, ' unknown'. Orwell felt pity for the microbe, he wondered how man had existed for decades in isolation, not being able and had not made any effort to interact with other members in his ecosystem. Orwell's break through thrilled him; he knew that with this discovery, he could also reach out to other members of the human ecosystem and eventually break the communication barrier between man and other members of his ecosystem, like the animals.

CHAPTER FIVE

A NEW FRIEND EMERGES

As days passed by, Orwell got to know more about the virus through his newly constructed device. This time around Orwell was prepared to take the adventure to the next level. As he walked into the laboratory early the next morning, he moved straight to the device and immediately typed in few sentences, asking the virus how she was feeling. And after about 5 minutes the virus replied saying " bored'. Orwell wondered what the virus meant by that and so he chatted back asking ' how?'.

'Your planet is inactive', replied the virus. Orwell was a bit confused by the virus's last response. Orwell wondered where the virus originated from and for how long she has existed on the planet?, Orwell chatted back again but this time more precise,' where do you come from then?', Orwell asked.

'™\\\¶Δ', the virus replied. Orwell couldn't understand the symbol; he concluded that it may be some kind of planetary code unknown to humans.

'You can't see me but I can see your cells', the virus replied. Orwell was astonished and so he replied-

'How do you mean'? At this point, the virus told Orwell that she came from a far star unknown to humans. The virus also let Orwell know that he can't see her physically unless the Earth's sun rays are thoroughly refined. The virus taught Orwell about the Zinoic rays, these are highly refined sun rays that can assist man see a lot of things invisible to his eyes. Ever since Orwell discovered the virus, this was actually the first time he had ever had a long heartily chat with her. Orwell looked at his watch, it was about 45 minutes left to close work and so he continued Chatting.

Zinoic Rays, the secret to Alien invasion on Earth

The virus let Orwell know that his planet (Earth) has been invaded by strange creatures and elements from far distant stars that are yet to be discovered by humans. These creatures and elements are invisible to the human eyes. Some of them do mingle with humans in several ways including, through human reproduction. The virus explained that most weird humans are products of these creatures and elements that had found their way into the human system. These creatures and elements are less aggresive if undictated but dangerous once they find out that they have been seen or dictated.

There are a number of factors that can prevent them from invading the human world, but none of those factors are accessible to humans because they are still primitive and archaic in comparison with what goes on in the universe. The virus made Orwell understand that there are a ton of rays in the universe as well as energy forms, all unknown to humans. Most of these Aliens frequent the Earth daily whiles some dwell on Earth permanently, trying out means to make humans compactible with their orientation.

CHAPTER SIX

ORWELL TAKES VIRUS TO HIS APARTMENT.

Orwell never wanted to set his eyes off his new friend, the virus, and so he decided to take her home to his apartment as soon as work closes. Orwell carefully dettached the tube from the device and lodged it safely at the back seat of his blue chivette car. As soon as he got home, he reserved a cool and calm spot for the virus and the device.

At exactly 8pm late in the evening Annie called in, it has been a while ever since they talked. She and Orwell were supposed to have a date by 9pm; however she decided to call in earlier to remind Orwell of their scheduled date. Orwell himself was busy sorting out few things with the virus, just to make sure that he gets the virus fully settled in her new home. As soon as he was done with that, he quickly left his apartment and headed towards Annie"s house to pick her up for the date. At Santos Hotel, located within town, Orwell and Annie sat directly opposite each other. They seem to be highly engulfed with each other in a romantic mood. Orwell held Annie's hand all through as they were chatting. Love was in the air but Orwell never had the chance to tell Annie of his strange relationship with a microorganism. He did know how Annie would feel about it if she gets to know of it. On the dance floor, two of them danced together to a known pop music that rolled for a while. Annie eventually decided to break the four months jinx by asking Orwell when they would be getting married. Orwell felt a bit embarrassed because he knew he hadn't proposed to Annie yet, and wondered how she is now talking of marriage? It was a hard nut to break for him. Orwell later told her in confidence that their wedding would likely come up next summer. Annie felt excited and gave him a long kiss with hugs as they danced to the music. Shortly after, Orwell drove Annie back to her house by exactly 10.55pm in the night. Orwell got home to his apartment, checked the virus and rushed to bed to have a rest.

CHAPTER SEVEN

THE DISCOVERY OF THE ZINOIC RAYS

Out of curiosity, Orwell started chatting with the virus as early as 3.13am in the morning. He wanted to know more about the zinoic rays and how it can be gotten. And the virus responded after 7 minutes with these texts, 'zinoic rays are integral part of the Earth's sun rays which can be extracted when sun rays are refined or filtered thoroughly by a powerful hyper-sensitive rays-filtering device', the virus replied.

Orwell was a bit satisfied but still wanted to know more, he then texted back asking, ' what components are required to construct the device?'. And the virus responded with a list of some highly technical items on the text screen. Then she added, ' The body of the tool should be able to trap sun rays and send them directly to the internal filtering components which filters and refines them at a high speed to produce zinoic rays. Zinoic rays are a million times more magnetic than the sun rays. When switched on, reveals highly magnetic bodies that normally can not be dictated by the sun rays,' the virus replied. Fascinated by the virus's theory, Orwell decided to consult his fellow colleague at the institute, Prof. Elin Marley, the following day. Prof Elin who happened to be a senior lecturer in physics department volunteered to help Orwell in his project, though Orwell never told him actually for what purpose the project was meant for. Both Orwell and Prof. Elin went into intensive research; they also consulted other professionals in related fields of studies.

Zinoic Rays, the secret to Alien invasion on Earth

After about two weeks of intense study and research, they sucesdfuly launched the first zinoic rays blazer. The device was designed to absorb sun rays and refine it to produce a distinctive version of the sun rays called the zinoic rays. Orwell jubilated over his achievement with the virus as soon as he got home. Later that evening, Orwell decided to launch the device by turning on the zinoic rays against the tube. Orwell marvelled as he saw the full image of the virus in the tube. Orwell went off the hook with excitement. He rushed out of his apartment with the device and switched it on on the streets of Novraska town.

What he saw scared him to death! He saw various shapes and manners of monstrous looking images that were busy walking about the town unseen by humans and living amongst humans for years. He quickly recalled moppy:s strange experience sometime ago and knew why the dog acted strangely because he was actually seeing these monstrous Aliens walking comfortably on the streets and establishing themselves on the human planet. They look like creatures from unknown stars and a billion miles away from Earth. As soon as the creatures saw the zinoic rays, they knew that a human has dictated them and so they became uncomfortable, wondering how it came to be. Most of them raged with fury and stomped towards Orwell. Orwell immediately switched off the device and ran for his life. Two huge and weird looking dogs appeared from no where and pursued Orwell with fury. Orwell suspected it could be those creatures in form of dogs and so he dived into his apartment and shot the door quickly. Orwell was so scared that he didn't know whether to continue or quit.

What he saw out there was horrifying, but before anything goes wrong he promised to let the whole world know about it. He checked on the virus and to his greatest shock, the virus was no longer there in the tube anymore. He wondered where on Earth the virus could have gone. The eerie situation brought goose pimples all over his body. Orwell quickly called both Annie and James on phone. He was afraid to step out, he waited until Annie arrived before James came.

Zinoic Rays, the secret to Alien invasion on Earth

Orwell felt like someone who has been chained, he narrated to his two friends how he met a strange virus and how the virus taught him about zinoic rays which got him into this mess. Right now the creatures are after his life and don't know the whereabout of the virus either. Annie suggested that the device be destroyed at once, but James had a different view.

He suggested that they call the corps and hand over the device to them. On a second thought, Annie decided to go out there with the device and see those creatures for herself. Orwell rejected Annie's decision and they argued over it for a while before Orwell let her go. James later volunteered to go with her and so they left Orwell alone in the apartment while they go searching for those creatures. At that time it was 10.45pm late into the night. Annie and James reached a particular spot on the lonely street and switched on the device. Two minutes after they switched it on, something terrible struck with a loud noise behind them and both Annie, James and the device went extinct, no where to be found again. Orwell waited fearfully till morning for their return, but could not find them or the device. At that point Orwell knew something unexplainable had happened to them. Who knows, he might be the on the line. Orwell stayed indoors for three days for the fear of those horrible invisible creatures out there. And on the fourth day, he quickly packed his things hoping never to return to Novraska town. He later settled in another town far away from Novraska. Uptill then, Orwell never knew the whereabouts of both the virus, Annie and James his friend. Orwell now became convinced that monsterous aliens from other worlds had invaded the Earth. May be some day in the future man will come to the awareness of these creatures and elements to either fight them or accept to be their next door neighbor. But for now, ' I have to run for my life' Orwell said to himself. .

Orwell relocated to Sinapoll, about 55 miles away from Novraska town. Sinapoll is a commercial town popularly known as a home for merchants. The town is always busy with different kinds of businesses all year round. Orwell after his experience two years ago at Novraska hired a two bedroom apartment in Marcos estate here in Sinapoll. Two of Orwell's neighbors, Mocova and Eloa both from Russian decent are both full time merchants of different commodities. None of them have time to visit each other, not to talk of having time to spend with a boring scientist like Orwell, who they refer to as a bore. Orwell on his own part never cared about what they feel about him. He was more concerned about the mysterious disappearance of his lover, Annie and colleague, James, two years ago. Ever since then, the families and relations of his two friends had been frantically searching and combing Novraska town in search of them.

As the searching intensified, some CID officers stormed Orwell's formal apartment in search of Orwell himself. They sought after Orwell after Mr. Lake, Orwell's former neighbor told the officers that he sometimes sees those two missing people in Orwell's apartment. And so, the officers concluded that Orwell may have clue to the whereabouts of the two missing persons and so, they tagged him as being wanted by the CID.

CHAPTER EIGHT

THE SEARCH

It got to Orwell's attention that he was being sought after by the CID and so he decided to report to their officer early in the morning of Tuesday. Sitting right in front of Sergeant Fabian, the district officer in charge of Novraska, Orwell summoned enough courage and narrated to the officer what actually transpired between the missing persons and him on the fateful night. After about two hours of intensive grilling of Orwell, sergeant Fabian decided to let him go but monitor him closely until they find something tangible to indict him.

No matter how hard Orwell tried to mingle and associate with his neighbors and friends at the office, he was still being tortured by the fear that one day those mysterious alien creatures would come after him to kill him. Orwell lived in fear of the unknown without letting anyone know about it. At Ralph Institute, a new lecturer was hired to replace the missing, Dr. James. The new lecturer who happens to be a lady of about 34 years from new Zealand took over class shortly after summer break and seems to be doing perfectly fine with the students.

Zinoic Rays, the secret to Alien invasion on Earth

One day, Elena, the new lecturer intercepted Orwell at the hall way after work and they spoke briefly for the first time ever since Elena came on board. On another occasion, Elena decided to take their cordial relationship to another level by inviting Orwell to her apartment over the weekends. Elena has been single for 34 years; she just moved into Novraska three months ago and has been living in one of the estates but not in Sarville estate.

On Friday evening after work, Orwell and Elena decided to date each other by 8pm later that day at Bouvo hotels in Novraska. At exactly 7.59pm, both of them were seated facing each other in what looks like a romantic mood. Orwell decided to tell Elena about his former relationship with Annie, who has been missing for two years. Elena surprisingly pretended not to have paid attention to that, instead she confessed to Orwell that she had known him even before they met each other for the first time at the hall way. Elena pulled out a shiny pendant to support her claims and showed it to Orwell.

Zinoic Rays, the secret to Alien invasion on Earth

At the sight of the shiny little pendant, Orwell sprang up from his seat, as he was able to recognize the pendant. Orwell looked at Elena with great intent; he wondered how Elena got to lay hold of a gift he gave to his microbe friend, the Virus, who mysterious vanished from the test tube on the same day that Annie and James also got missing two years ago. Orwell summoned up enough guts and leaned closer to Elena and asked her how she got the pendant.

When Elena saw that Orwell was creating a scene in the hotel hall, she quickly stood up and took Orwell by his left arm and walked him outside the hall to the balcony of the hotel. Elena tried everything possible to calm down his nervous state by promising him that she will tell him all that he wants to know by the next day, Saturday, exactly 6pm in the evening time at her apartment in leopard estate, Novraska. Orwell at this point feared that Elena may have something to do with those mysterious creatures two years ago. He felt a bit unsecured with Elena, but somehow managed to keep calm until the next day. He dropped Elena off at her apartment and shot the door of his car without uttering a single word to her, and he drove off immediately.

CHAPTER NINE

THE TRUTH

At exactly 6.02pm the next day, Orwell arrived Elena's apartment. Elena was already waiting for his coming at her lounge. Immediately the doorbell rang, she walked towards the front door and gently opened the door, knowing fully well that it could be Orwell himself. Orwell walked and sat down in one of the seats in the parlour. Elena attempted offering him a drink, but he turned her offer down. Orwell seems to be a bit nervous. Elena eventually settled in one of the seats very close to that of Orwell's, they looked at each other like a ghost. Finally Orwell decided to break lose the ten minutes suspense by throwing in the first question at Elena. He looked straight at Elena's eye balls and asked her for the second time, 'where did you get the pendants?

Elena stressed back her seat a little and replied with a gentle and quite tone,' I am the Virus in the test tube', she said. Orwell looked intently at her the third time and asked, 'how?. Then Elena stood up to her full length, geared to let the cat out of the bag. She walked towards Orwell and held his hand gently over her cheek,' you gave me this precious pendant when I was in the tube, and I prayed so hard to behold your face and hold your hands one day. Now the time has come to tell you the truth'. Orwell began to sweat, he replied, 'what truth?. 'I am that microbe friend you discovered at the bark of an ancient tree many years back, and you put me in a test tube'. Elena said. Orwell looked at her with awe and said, 'how then have you become human'.

Zinoic Rays, the secret to Alien invasion on Earth

'Because I loved you and didn't want to lose you to Annie, and so I had to escape from the tube and transform into human'.

Orwell stood up from his chair and walked towards her, he held her hands and said to her,' who then are you and where can I find Annie and James?'

Elena held his hands tight and replied back,' you will never see then until 300million years, both of them are now stars of their own galaxies. They are no longer humans; they are now new forms of energy'. Orwell replied,' how?'.

'You sun produces millions of different rays every 500 hours, some of these rays end up merging or pairing with one another to produce a powerful ball of new emerging energy. It eventually buds off to a distant location to become a star of its own. The new ball of energy then multiplies itself to form a galaxy of same form of energy. Sometimes these baby energy forms explore other galaxies of different energy forms. These energy forms are what you humans call alien creatures or UFO. Your galaxy is unique because it is the source of all the rays, it produce countless versions rays. But unlike other galaxies, your galaxy inhabits organic creatures like you humans. I saw some of those archaic crafts of yours scattered all over your galaxy, what really are they for?'. Elena asked Orwell.

Orwell replied with enthusiasm, 'you mean the space ships and rocket launchers on the orbit?' Elena burst into laughter, and said to him,' how archaic'.

Zinoic Rays, the secret to Alien invasion on Earth

Elena continued, but this time more direct and specific to her point, she said, 'you humans need to protect your galaxy and race, there are greater forces out there that are gearing up to submerge you galaxy and take over. If care is not taken, the human race will go extinct in a short time'. Orwell became a bit confused and at the same time worried, he then said to Elena,' how then can man protect his galaxy and race since he can't see who he is battling with?'.

'You can save your galaxy and race if you are willing to explore and learn'. Elena replied looking Orwell straight in the eye.

'I am willing, if that will save of galaxy and race?'. Orwell replied. Elena seeing how willing and curious Orwell was, she dipped her hands inside her jean pocket and brought out what looked like a small bluish cork and handed it over to Orwell. She now said to him, 'take this and swallow it at exactly 2am. This will decompose your organic body and transform you into Epiglosisis energy form which I also belong to in my galaxy'. Orwell quickly took the object from her hand and examined for a while, and then he said,' where is your galaxy?' Elena replied.'4million miles from Earth'. Orwell thanked her for her concern and at the same time wasn't really sure of himself. Orwell left Elena's house feeling like an alien creature itself. He arrived his apartment late in the night and sat down on the sofa thinking over what transpired between him and Elena. He paced about his parlor several times, thinking over things he will leave behind if he leaves, his friends and family?, his work at the institute?, his neighbors?, and even his enemies?. And Orwell thought to himself,' if I don't do this, then our galaxy and the human race including his loved ones will be in great danger?' Orwell made up his mind and went to shower.

Zinoic Rays, the secret to Alien invasion on Earth

And at exactly 2am, a sudden bright flash of light came out from Orwell's door and disappeared immediately. All Orwell's neighbors saw the bright flash of light but never knew where it came from. From that moment on, Orwell became history.

www.ingramcontent.com/pod-product-compliance
Lightning Source LLC
Chambersburg PA
CBHW071204220526
45468CB00003B/1158